Innovative Manufacturing 2018

Topical Conference at the 2018 AIChE Spring Meeting and
14th Global Congress on Process Safety

Orlando, Florida, USA
22 – 25 April 2018

ISBN: 978-1-5108-6434-4

Printed from e-media with permission by:

Curran Associates, Inc.
57 Morehouse Lane
Red Hook, NY 12571

Some format issues inherent in the e-media version may also appear in this print version.

Printed by Curran Associates, Inc. (2018)

For permission requests, please contact AIChE
at the address below.

AIChE
120 Wall Street, FL 23
New York, NY 10005-4020

Phone: (800) 242-4363
Fax: (203) 775-5177

www.aiche.org

Additional copies of this publication are available from:

Curran Associates, Inc.
57 Morehouse Lane
Red Hook, NY 12571 USA
Phone: 845-758-0400
Fax: 845-758-2633
Email: curran@proceedings.com
Web: www.proceedings.com

TABLE OF CONTENTS

Some abstracts in this book refer to images that were not made available by the publisher

514236 Integration of Smart Manufacturing in the Advancement of Polymer Processing

Abstract:

In lieu of improving efficiency and embracing smart manufacturing technologies, the U.S. Department of Energy (DOE) has instituted the Clean Energy Smart Manufacturing Innovation Institute (CESMII), with Texas A&M Energy Institute leading the Gulf Coast Regional Manufacturing Center. The goals of CESMII are to (1) demonstrate at least 15% improvement in energy efficiency in real plant demonstration projects within five years, supporting a later goal of 50% improvement in energy productivity in 10 years; and (2) reduce the cost of deploying Smart Manufacturing initiatives to processing plant by 50% through the development tools and technologies. These includes developing platforms that enables machine-to-plant-to-enterprise real time sensing, instrumentation, monitoring, optimization, and control of energy. The idea of Smart Manufacturing integrates our expertise in high fidelity or advanced modeling of industrial operation, big data analytics, real time optimization and leverages high performance computing to achieve a seamless development of optimal processes and energy reduction.

The most recent advances in sampling technology and sensor development has enabled continuous online monitoring of polymerization reactions, leading to an opportunity for the application of advanced controllers. In this work, we develop model based advanced control strategies for batch polyolefin reactions via the PARametric Optimization and Control (PAROC) framework. Based on accurate linear representations of an initial high fidelity model, we formulate a Model Predictive Control (MPC) scheme that delivers a trajectory tracking problem. We solve the multi-parametric counterpart of the MPC problem (mpMPC) to determine the explicit expressions of the optimal control actions. Hence, the online computational burden of the optimization problem is reduced to a simple lookup table and affine function evaluation problem. Such an alleviation in the computational cost and ease of application allows for an opportunity for the integration of the control strategies in the Smart Manufacturing cloud platform to be used by industrial partners.

513861 Smart Manufacturing: A Must to Remain Competitive

Abstract:

Rapidly evolving technologies and products are posing enormous challenges and opportunities for chemical manufacturers. Meeting these challenges is a bare minimum for surviving in todayâ€™s dynamic global markets. Whereas to be competitive a manufacturer must be able to transform each and every rising opportunity into cash or strategic benefits.

Smart Manufacturing offers a multitude of capabilities to competitively deal with the ever increasing challenges and opportunities of chemical manufacturing.

Smart Manufacturing is a set of enabling technologies, methodologies and tools that can be deployed to improve the core business specific technologies, operations and work processes

The components of Smart Manufacturing are generic and are usually applicable to a number of different manufacturing processes.

In this presentation we will elaborate on main components of Smart Manufacturing in the context of chemical manufacturing. Proper deployment of tools such as big data analytics, process control and optimization, real-time fault detection & diagnosis, and predictive maintenance will also be described. A strategy to adopt Smart manufacturing will also be suggested.

513097 Development of an Advanced Control Philosophy for Dry Gas Combined Steam Reforming, to Operate the Plant Efficiently during Feed Composition Changes

Abstract:

Steam reforming of natural gas is one of the very important processes in the world now, as hydrogen, a key product of the reforming, continues to be in high demand for various petrochemical and chemical industries. As the gas reserves are depleting in the world, existing reforming plants are open to use natural gas from different sources. Different natural gas sources offer different compositions in the gas. This wide range of feed natural gas composition has significant effect on the performance of the steam reformers. In order to operate the reforming process at optimum conditions, precise control of reformer temperature, and steam intake are very crucial. Disturbances in the feed gas create instability in the reforming process. Furthermore, some steam reformers are operated with carbon dioxide added to the reformer feed, to have a lower hydrogen to carbon mole ratio, as required for downstream methanol or plasticizer alcohols processes. These dry gas combined steam reformers are operated at higher temperatures and lower steam to carbon ratios than that of pure steam reformers. Hence, frequent feed composition disturbances imply greater in-stability in the dry gas combined steam reformers in terms of reformer performance, reformer temperature, methane slip, fuel gas consumption and excess air in reformer box.

The existing control philosophy of the reformers cannot cope with the feed disturbances and can be effected by big swings in the reformer temperature, methane slip, fuel gas and steam & carbon dioxide flows in case of dry gas combined steam reformer. These disturbances may also have adverse effects on the plant efficiency, capacity, safety and reliability, and the plant integrity. In order to have a stable operation and cope with these disturbances, most of the existing plants are opting for advanced process control possibilities like Aspen DMC+ etc

In the paper, a new feed forward control strategy is described that can control the dry gas combined steam reforming process efficiently and effectively for a large composition variations. The basic methodology of the new control strategy is to control the steam reforming key parameters like reformer temperature, stream flow, process air, CO2 flow, and Fuel gas flow, based on the carbon count of the natural gas in a feed forward way. Additionally, a small correction to the individual manipulators is done in a feedback control philosophy. This feed-forward & feedback control philosophy enables the operating plant to react to the disturbances quickly and efficiently. Hence, the plant performance disorder is kept minimum and huge swings in the parameters are avoided during the feed disturbances. Also, with this new control strategy, the variations in the plant parameters are reduced and stable and steady production of the hydrogen can save the energy about 0.5 GJ/ton of H2 produced from the reforming process. More than the energy, the process operates more steadily and downstream processes that depend on the reforming process operate steadily as well. This developed strategy is tested and implemented and has improved the plant reliability and integrity.

513495 Smart Systems for the Process Industries

Abstract:

ABSTRACT for AIChE 2018 Spring Meeting

Smart Systems for the Process Industries

Samvith Rao and James C Cross III

The MathWorks, Inc.

samvith.rao@mathworks.com

508-647-4487

An unprecedented and increasing level of device and computing capability and connectedness is intensifying competition and paving the way for new business opportunities. The challenge for existing companies and new ventures alike is to: determine what information is important; decide how to collect and process it; and conceive, develop, and offer new products and services that deliver value from it. This is the aspiration of the Industrial Internet of Things (IIoT), and Industry 4.0.

Specifically, the proliferation of inexpensive sensors, wireless communications, big data and cloud infrastructure, machine and deep learning algorithms, and ever-faster high-performance computing platforms is ushering in a new era of intelligent and autonomous systems, including self-driving cars, smart industrial systems, and automated warehouses. Moreover, the confluence of these technologies has given rise to a new type of industrial asset – a virtual, real-time replica of a companion physical system, the so-called *digital twin* – that can be leveraged for a wide range of operational benefits, including process optimization, predictive maintenance, fault detection, model-based control, and dynamic pricing.

In this talk we describe the individual enabling technologies, present examples of how they are being developed or used commercially, and offer perspectives on their adoption and potential within the process sector.

512516 Smart Damper Design Provides Better Control of Fired Heaters

Abstract:

Fired heaters are major consumers of energy in the Refining and Petrochemical Industries. In most of the natural draft fired heaters, one of the important parameters that is controlled by the operators is the furnace draft using stack dampers. Historically stack damper quality in the industry is very poor. The current design of the stack dampers is inadequate to control the draft effectively. As a result most of the stack dampers are left fully open in the industry. The average draft in the fired heaters is maintained at almost 3-4 times the value recommended causing a lot of tramp air in the system.

In this paper; we are going to talk about how our patent pending Smart Stack Dampers can ensure proper draft control in fired heaters. We have made several improvements in the design and construction of the dampers. Smart dampers are designed to be very reliable and operator friendly. They provide a much better control of the draft. Smart dampers can be configured for automatic or manual draft control of the heater

513860 Application Centralized Information Technology in Process Design Package Development

Abstract:

Smart Manufacturing solution is built on information technology, researchersâ€™ ingenuity to create innovative product, continuously analyzing market and manufacturing data to discover the market needs and signal problems impacting your costs, quality, and production of the existing technology. Centralized database technology with advanced simulation are keys to quickly delivery of the new products reliably, safely, and environmental friendly.

This presentation provide an introduction for applying centralized information technology in process design package development (1) evaluating new innovative technology for its reliability, product profitability, global market potential, sustainability and technology; (2) Aligning proven and conventional technology to form an integrated commercial process based on commercial database; (3) develop the process design package and expand the process globally in a short timeframe.

NOTES

NOTES

NOTES

NOTES

AUTHOR INDEX